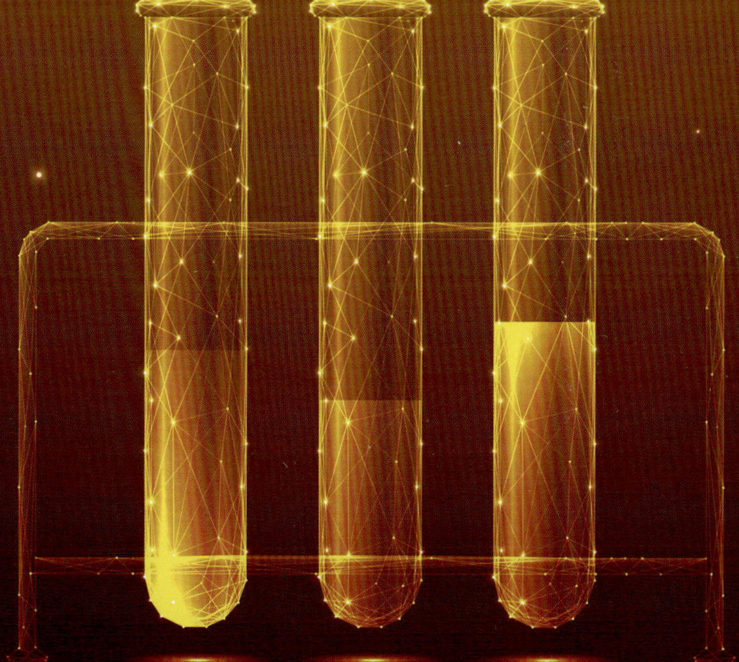

THE pH SCALE

BY MARIE ROESSER

Please visit our website, www.enslow.com. For a free color catalog of all our high-quality books, call toll free 1-800-398-2504 or fax 1-877-980-4454.

Library of Congress Cataloging-in-Publication Data

Names: Roesser, Marie, author.
Title: The pH scale / Marie Roesser.
Description: New York : Enslow Publishing, [2026] | Series: Chemistry in review | Includes index.
Identifiers: LCCN 2024038689 (print) | LCCN 2024038690 (ebook) | ISBN 9781978543003 (library binding) | ISBN 9781978542990 (paperback) | ISBN 9781978543010 (ebook)
Subjects: LCSH: Hydrogen-ion concentration–Juvenile literature. | Acids–Juvenile literature. | Bases (Chemistry)–Juvenile literature.
Classification: LCC QD562.H93 R64 2026 (print) | LCC QD562.H93 (ebook) | DDC 541/.3728–dc23/eng/20240929
LC record available at https://lccn.loc.gov/2024038689
LC ebook record available at https://lccn.loc.gov/2024038690

Published in 2026 by
Enslow Publishing
2544 Clinton Street
Buffalo, NY 14224

Portions of this work were originally authored by Kennon O'Mara and published as *The pH Scale* (A Look at Chemistry). All new material in this edition is authored by Marie Roesser.

Designer: Claire Zimmermann
Editor: Therese Shea

Photo credits: Cover, p. 1 (main image) AntonKhrupinArt/Shutterstock.com; series art (molecule header image) jijomathaidesigners/Shutterstock.com; p. 5 Charles Brutlag/Shutterstock.com; p. 7 Robert Kneschke/Shutterstock.com; p. 9 PeopleImages.com - Yuri A/Shutterstock.com; p. 11 (top) chemical industry/Shutterstock.com, (bottom) Artem Stepanov/Shutterstock.com; p. 13 hikmet2016/Shutterstock.com; p. 25 PawelKacperek/Shutterstock.com; p. 29 Nataliia Melnychuk/Shutterstock.com; p. 30 BlueRingMedia/Shutterstock.com.

Printed in China

CPSIA compliance information: Batch #QSENS26: For further information contact Enslow Publishing, at 1-800-398-2504.

CONTENTS

Words in the glossary appear in **bold** the first time they are used in the text.

IT'S CHEMISTRY!

Do you like lemonade? Did you ever wonder why it's sour? That taste has to do with natural **chemicals** called acids. Acids have many **properties**. Sour taste is just one. Chemicals with opposite properties are called bases. We can measure how acidic or basic something is with the pH **scale**.

LEARN MORE

Some acids and bases are natural. Some are made by people.

ALL ABOUT ACIDS AND BASES

Acids can be weak or strong. Weak acids are in many foods and drinks. Citric acid is a colorless weak acid in lemons. Acids feel a bit rough if you touch them. But strong acids are too strong to touch! Sulfuric acid is used in car **batteries**. It burns skin.

LEARN MORE

Lactic acid is found in yogurt and pickles.

Bases can be weak or strong too. But bases taste bitter, not sour. They feel soapy when they're wet. Baking soda, or sodium bicarbonate, is a weak base used in many kinds of foods—such as cookies and cakes.

LEARN MORE

When a base is mixed with an acid, gases form. A mix of baking soda (base) and vinegar (acid) is often used to make volcano models!

Strong bases are dangerous, just as strong acids are. They can burn skin easily. Lye, also called sodium hydroxide, is a strong base. It can break down animal and plant parts. But it's also useful. People make soap and paper with lye.

LEARN MORE

A base that **dissolves** in water is called an alkali. Lye is an alkali.

THE pH SCALE

The pH scale can tell us how acidic or basic something is. The scale uses numbers, from 0 to 14. **Substances** with a pH less than 7 are acidic. Substances with a pH greater than 7 are basic.

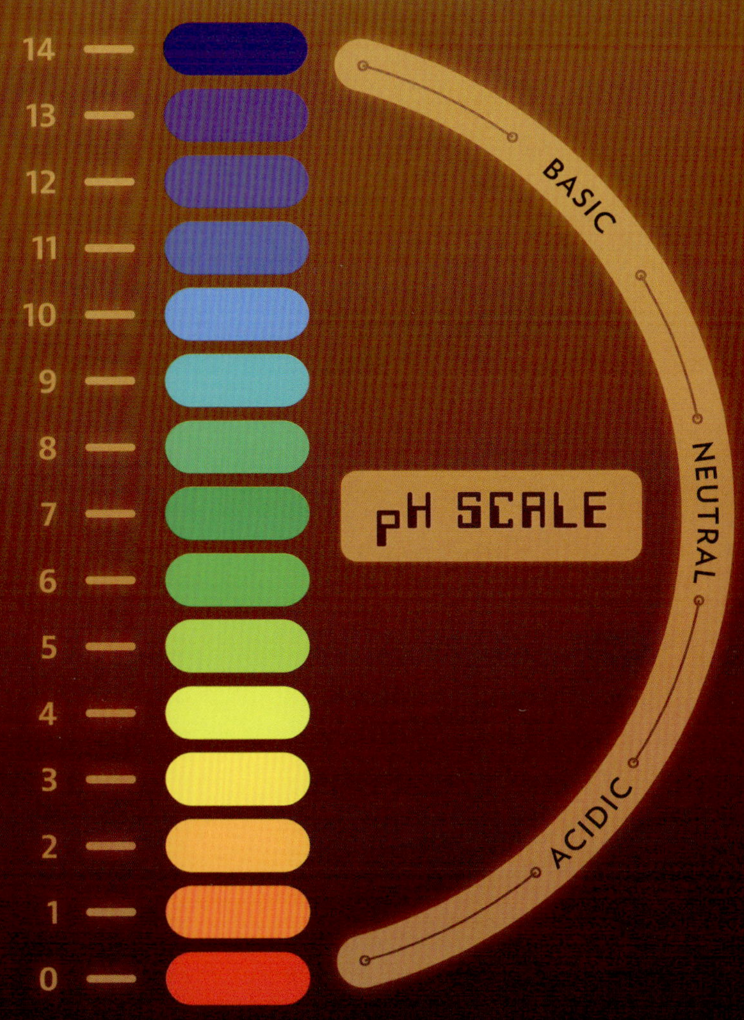

LEARN MORE

The lower the pH number, the more acidic something is. The higher the number, the more basic.

ATOM INVESTIGATION

What *is* pH? It's a measurement of how much of a certain kind of **atom** is present. Atoms are the smallest part of an element. Elements are building blocks of matter. Every element is made of a **unique** kind of atom. Hydrogen and oxygen are elements.

HYDROGEN ATOM

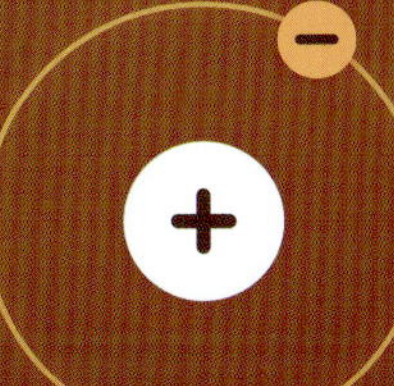

proton

electron

OXYGEN ATOM

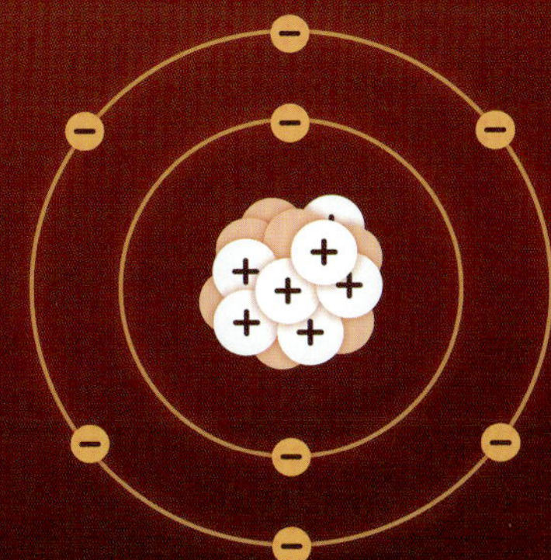

8 protons

8 electrons

8 neutrons

LEARN MORE

Atoms are made of **particles**. An atom's center, or nucleus, contains protons and neutrons. Electrons travel around the nucleus.

Each proton in an atom has a positive **electric** charge. Each electron has a negative charge. An atom usually has an equal number of protons and electrons, so the charges balance. An atom with no electric charge is called neutral.

CARBON ATOM

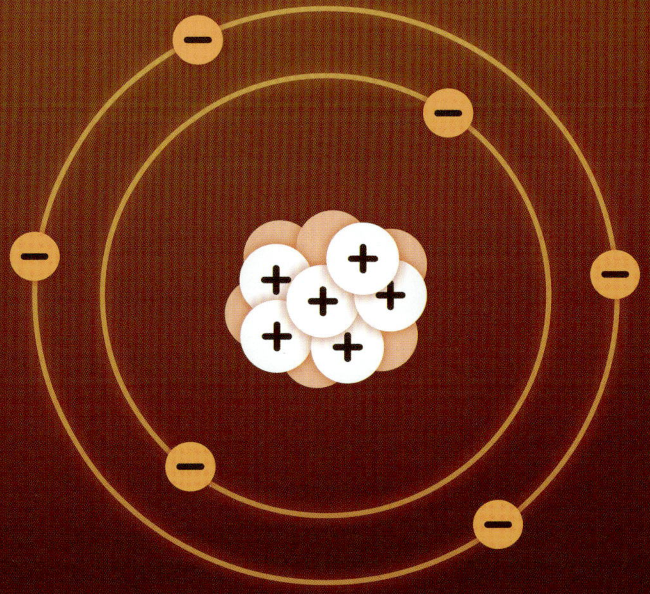

+ 6 protons − 6 electrons  6 neutrons

LEARN MORE

This is a model of a neutral carbon atom. Neutrons have no electric charge.

IS IT AN ION?

In nature, atoms can gain and lose electrons. They **react** with chemicals and crash into other atoms. An atom that has gained or lost an electron is called an ion. Its overall charge is no longer in balance. It's positively or negatively charged.

THREE KINDS OF HYDROGEN ATOMS

LEARN MORE

A positively charged ion is called a cation. A negatively charged ion is called an anion.

An acidic substance has a larger amount of hydrogen ions than hydroxide ions. A basic substance has a larger amount of hydroxide ions than hydrogen ions. A substance that has equal amounts of these kinds of ions is called neutral.

ACID

hydrogen chloride

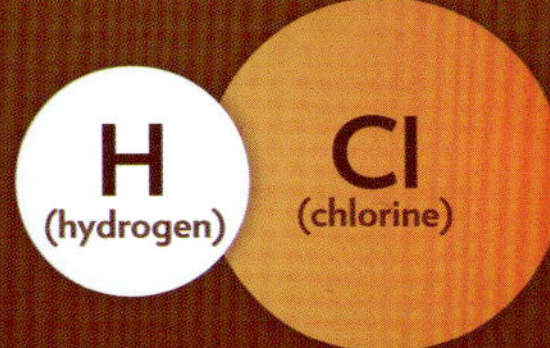

chloride ion

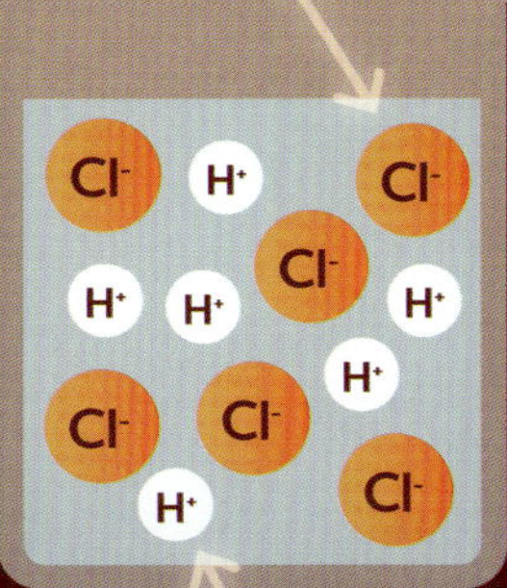

hydrogen ion

BASE

sodium hydroxide

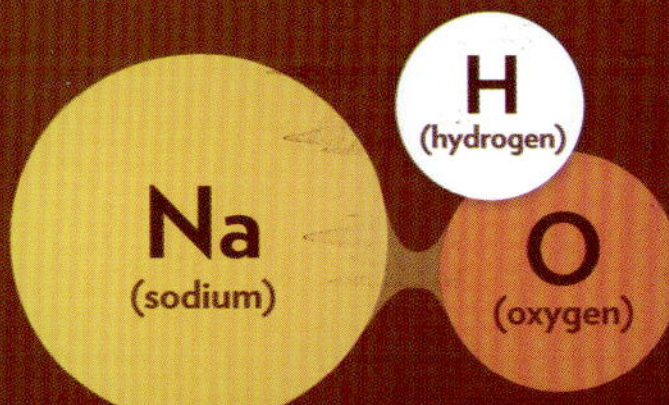

hydroxide ion

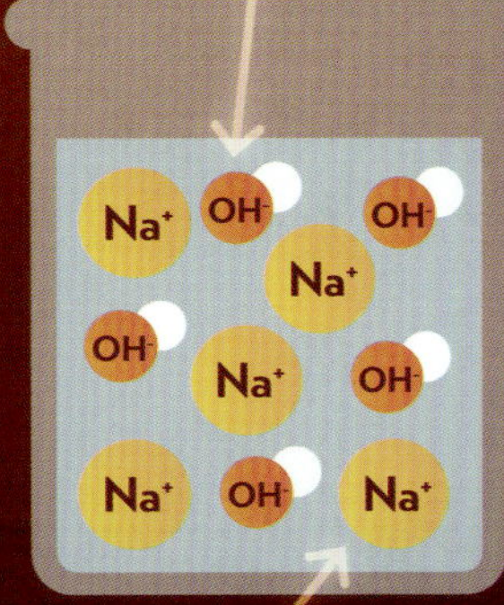

sodium ion

LEARN MORE

Hydroxide is made up of hydrogen and oxygen atoms. It's a molecule, which is a group of chemically bonded atoms.

TEST IT!

We can't see hydrogen ions and hydroxide ions. To test whether a substance is an acid or a base, scientists use indicators. Indicators indicate, or show, the presence of an acid or base, often with a color change. A well-known indicator is litmus paper.

LITMUS TEST

Blue litmus paper turns red in an acid.

Red litmus paper turns blue in a base.

LEARN MORE

To test the pH of a solid, scientists mix it with water that has no ions. This mixture is called a **solution**. The color change we see on the litmus paper is a chemical reaction!

Scientists can test for the strength of an acid or base too. An indicator called universal, or alkacid, indicator paper can turn many colors in a solution. The color the paper turns is compared to colors on a pH scale chart, from 1 to 14.

LEARN MORE

The term pH is an abbreviation, or shortened form of a word or words. The "p" in pH stands for *potenz*, which is the German word for "power." "H" is short for "hydrogen."

WHAT ABOUT WATER?

Pure water has an equal amount of hydrogen ions and hydroxide ions. If you tested it with universal indicator paper, it would turn green. This means pure water has a pH of 7. Solutions with a pH of 7 are neutral. They are neither acidic nor basic.

UNIVERSAL INDICATOR PAPER

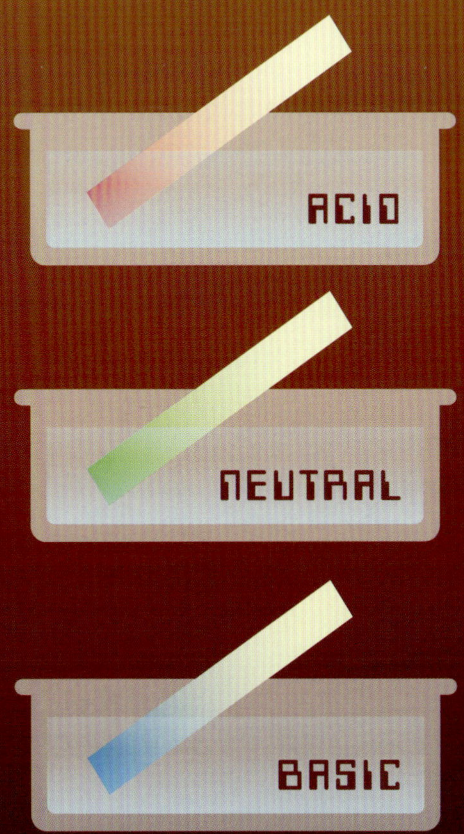

LEARN MORE

If you blew into pure water, the carbon dioxide in your breath would make the solution acidic, lowering its pH!

WHY WE NEED TO KNOW

Crops need soil of a certain pH. For example, strawberries need acidic soil of about pH 5 to pH 6.5. Farmers can add an acid or base to the soil if needed. Scientists test our drinking water's pH to make sure it's healthy. The pH scale is helpful in many ways!

LEARN MORE

Human blood is usually a bit basic, with a pH of around 7.4.

ACIDS AND BASES IN YOUR WORLD

GLOSSARY

atom: The smallest bit of matter that has all the properties of a chemical element.

battery: A device that turns chemical energy into electricity.

chemical: Matter that can be mixed with other matter to cause changes.

dissolve: To mix with a liquid and become part of the liquid.

electric: Having to do with the flow of charged particles.

particle: A very small piece of something.

property: A quality or feature of something.

react: To change after coming in contact with another substance.

scale: A range of numbers used to show the size or strength of something.

solution: A liquid in which something has been dissolved.

substance: A certain kind of matter.

unique: One of a kind.

FOR MORE INFORMATION

BOOKS

Adams, William D. *Acids, Bases, and Salts.* Chicago, IL: World Book, 2023.

Wood, Kevin. *The Empty Science Lab: Solve Your Way Out!* New York, NY: Gareth Stevens Publishing, 2023.

WEBSITE

Chemistry for Kids: Acids and Bases
mail.ducksters.com/science/acids_and_bases.php
Check out this review of key chemistry ideas.

INDEX